The Future of SEO: A World Where ChatGPT And Similar AI Technologies Replace Search Engines

BY BRIAN C EBY

Table of Contents

Introduction

The evolution of search engine optimization (SEO) has always been closely tied to the technological advancements of the internet. From the early days of keyword stuffing to Google's sophisticated algorithms, SEO has adapted to every shift in the digital landscape. However, a new and potentially disruptive technology has emerged: advanced AI models like ChatGPT. These AI systems have the potential to alter how people search for and consume information online fundamentally. This Booklet briefly, yet succinctly, explores the hypothetical scenario where AI models like ChatGPT significantly reduce or even eliminate the use of traditional search engines, examining the potential future of SEO in such a landscape.

CHAPTER

ONE

The Rise of AI in Information Retrieval

1.1 The Evolution of Search Engines

Search engines have undergone significant evolution since their inception. Early search engines, such as AltaVista and Yahoo!, relied on simple algorithms that indexed pages based primarily on keyword frequency. Google revolutionized the field with its PageRank algorithm, which considered the quality and quantity of backlinks to determine the relevance of web pages. Over time, Google's algorithm grew increasingly sophisticated, incorporating machine learning, natural language processing (NLP), and other advanced techniques to understand user intent better and provide more accurate search results.

However, despite these advancements, traditional search engines still rely heavily on a model where users input a query and receive a list of links that may or may not answer their questions directly. This model has limitations, particularly regarding complex queries requiring nuanced understanding or synthesizing information from multiple sources.

The rise of search engine giants like Google, Baidu, and Bing has made search engines the gatekeepers of online information. They control what users see and how businesses reach their audiences. The algorithms these companies use to rank content are guarded secrets, leading to an entire industry built around understanding and optimizing for these algorithms—SEO.

1.2 The Emergence of ChatGPT and AI-Powered Information Retrieval

ChatGPT and similar AI models represent a new paradigm in information retrieval. These AI systems are designed to understand and generate human-like text, enabling them to engage in conversational interactions with users. Unlike traditional search engines, which provide a list of links, AI models like ChatGPT can deliver direct answers, explanations, and even advice in response to user queries.

The potential advantages of AI-powered information retrieval are clear: users can receive more personalized, contextually relevant information without sifting through multiple search results. Moreover, AI systems can provide detailed explanations and engage in follow-up questions, offering a more interactive and dynamic experience than traditional search engines.

One significant advantage of AI models over traditional search engines is their ability to synthesize information from various sources and present it cohesively. Instead of navigating through multiple websites, users receive comprehensive answers in one interaction, saving time and effort. However, this also centralizes information distribution, raising concerns about content diversity and bias.

In addition, AI models like ChatGPT can process and analyze large volumes of data in real-time, allowing them to stay up-to-date with the latest information and trends. This real-time processing capability means that AI-driven information retrieval systems could offer more accurate and timely responses than traditional search engines, which rely on periodic index updates.

CHAPTER TWO

The Impact on SEO

2.1 The Decline of Traditional Search Engines

If AI models like ChatGPT become the primary information retrieval method, traditional search engines' role would likely diminish. This shift could have profound implications for SEO, which has long been centered around optimizing content for search engine algorithms. In a world where users turn to AI for answers rather than searching for links, the metrics that currently drive SEO—such as keyword rankings, backlinks, and click-through rates—might lose their relevance.

This decline in the use of traditional search engines could lead to a significant restructuring of the digital economy. Companies that rely heavily on search engine traffic for their business models must find new ways to reach their audiences. This could involve shifting their focus from search engine rankings to building direct consumer relationships through personalized content and engagement strategies.

The decline of traditional search engines would also have significant implications for advertising. A substantial portion of online advertising revenue is generated through search engine advertising platforms like Google Ads. If users increasingly turn to AI models like ChatGPT for their information needs, advertisers may need to find new platforms and strategies to reach their target audiences. This could lead to new advertising models better suited to AI-driven environments.

2.2 The Shift in SEO Strategies

As the focus shifts from search engines to AI-driven information retrieval, SEO strategies will need to adapt. Here are some potential changes:

- **Content Optimization for AI**: Instead of optimizing content for search engine algorithms, content creators would need to focus on optimizing for AI models. This could involve ensuring that content is easily understandable by AI, written in a conversational tone, and structured in a way that allows AI to extract relevant information effectively.

Optimizing AI will require a deep understanding of how AI models process and interpret information. This might involve creating content structured to align with the training data used by AI models and experimenting with different content formats to determine which are most effective for AI-driven information retrieval.

- **Conversational Content**: Since AI models like ChatGPT excel in conversational interactions, there may be a greater emphasis on creating content miming natural human conversation. This could lead to a rise in content that is more engaging, personalized, and tailored to specific user needs.

The shift toward conversational content will likely require businesses to adopt a more personalized and customer-centric approach to content creation. Rather than simply pushing out generic content, companies will need to engage with their audiences deeper, using AI to deliver content that resonates with individual users based on their preferences, behavior, and past interactions.

- **Focus on Expertise and Authority**: AI models rely on large datasets to generate responses, meaning content from authoritative and reputable sources will likely be prioritized. This could make E-A-T (Expertise, Authoritativeness, Trustworthiness) principles even more critical in the future of SEO.

As AI models become more sophisticated in their ability to evaluate the credibility of sources, businesses will need to focus on building their reputations as trusted authorities in their industries. This could involve investing in thought leadership, creating high-quality content demonstrating expertise, and engaging with industry influencers to build credibility and trust.

- **Diversification of Content Formats**: AI models can process and generate text but are also increasingly capable of handling other content formats, such as audio and video. As a result, SEO may need to encompass a broader range of content types, emphasizing creating multimedia content that AI can understand and utilize.

Diversifying content formats will be essential for reaching audiences across different platforms and devices. This could involve creating content optimized for voice search, developing interactive experiences that engage users in real-time, and leveraging emerging technologies like augmented reality (AR) and virtual reality (VR) to create immersive content experiences.

2.3 The Role of Keywords in an AI-Driven World

In traditional SEO, keywords have been the cornerstone of content optimization. However, in a scenario where AI models like ChatGPT dominate information retrieval, the role of keywords may change significantly. Rather than focusing on specific keywords, content creators must prioritize their content's overall context and relevance.

AI models are designed to understand natural language, meaning they can interpret the meaning behind user queries rather than rely solely on specific keywords. As a result, the emphasis may shift from keyword density to the quality and clarity of the content. Content that directly answers questions provides detailed explanations and aligns with user intent could become more important than targeting specific keywords.

Moreover, AI models could make searching more accessible for users using more natural and complex language. This could lead to a decline in the importance of exact-match keywords and an increase in the importance of long-tail keywords and phrases that reflect how people naturally speak and think. As a result, SEO strategies may need to evolve to focus on creating content that is more aligned with user intent and less reliant on specific keywords.

This shift could also change the tools and techniques used for keyword research and content optimization. Traditional keyword research tools may become less relevant as AI-driven information retrieval systems become more sophisticated. Instead, content creators may need AI tools to analyze user intent and generate content recommendations based on natural language processing.

CHAPTER THREE

The Evolution of Content Creation

3.1 AI as a Content Creator

One of the most significant implications of AI models like ChatGPT is their ability to generate content autonomously. This capability can disrupt the content creation industry, as AI could produce high-quality content at scale, reducing the need for human writers. However, this shift also presents new opportunities for content creators who can leverage AI to enhance their work.

Using AI in content creation could significantly increase the volume of online content. This could result in a more competitive landscape, where businesses must differentiate themselves by creating content that stands out in quality, originality, and relevance. While AI can generate content quickly and efficiently, human creativity and expertise will still be essential for creating content that resonates with audiences on an emotional and intellectual level.

3.2 The Role of Human Creativity and Expertise

In a future where AI plays a central role in content creation, human writers will be more critical than ever. They will need to focus on tasks that require creativity, critical thinking, and specialized knowledge. These tasks, such as creating unique, high-quality content that AI cannot easily replicate, will set human writers apart in the AI era.

Human writers ensure that AI-generated content is accurate, ethical, and aligned with brand values. While AI can generate content based on data and algorithms, it is human judgment that provides the context and ethical considerations necessary to create trustworthy and responsible content.

Moreover, human writers can leverage AI to enhance their creativity and productivity. For example, AI can generate ideas, conduct research, and create initial drafts, allowing human writers to focus on refining and polishing the content. This collaborative approach could lead to a more efficient content creation process, where AI and human writers work together to produce scale-quality content.

3.3 Collaboration Between AI and Human Writers

Rather than replacing human writers, AI could be a powerful tool to augment their capabilities. For example, AI could generate initial drafts, conduct research, or provide insights into trending topics. Human writers could refine and enhance this content, adding their unique perspectives and expertise. This collaborative approach could lead to more efficient content creation processes while maintaining the quality and creativity that human writers provide.

Collaboration between AI and human writers could also lead to the creation of new content formats and experiences. For example, AI could generate real-time interactive content that adapts to user preferences. At the same time, human writers could focus on creating the narrative structure and ensuring that the content aligns with brand messaging.

Additionally, AI could optimize content for different platforms and audiences, allowing human writers to focus on creating content that resonates with specific target markets. This could involve using AI to analyze user data and generate content recommendations tailored to individual preferences and behaviors.

3.4 The Importance of Originality and Authenticity

As AI-generated content becomes more prevalent, the value of originality and authenticity in content creation will likely increase.

Readers and users may become more discerning, seeking content that offers unique insights, personal experiences, and a genuine voice. This resurgence in the importance of brand storytelling and content that resonates emotionally is an inspiring opportunity for human writers to leverage their distinct advantage over AI.

In a world where AI can generate content at scale, businesses must urgently create original and authentic content to stand out. This could involve leveraging brand values, customer stories, and unique perspectives to create content that resonates with audiences on a deeper level.

Moreover, businesses may need to be more transparent about using AI in content creation, mainly to build trust with their audiences. This commitment to transparency, including clearly labeling AI-generated content and providing context about how AI is used to enhance the content creation, reassures audiences about ethical content creation practices and fosters a sense of trust.

CHAPTER
FOUR

The Role of User Experience in AI-Driven SEO

4.1 The Importance of Contextual Understanding

In an AI-driven world, user experience (UX) will play a crucial role in determining the success of SEO strategies. AI models like ChatGPT are designed to understand the context of user queries and provide relevant, tailored responses. As a result, content designed with user intent in mind and that offers a seamless and intuitive experience will likely perform better in AI-driven information retrieval systems.

Content creators and marketers must create content that aligns with user intent and provides a positive user experience. This could involve optimizing content for different devices and platforms, creating intuitive navigation structures, and ensuring that content is easy to read and understand.

Moreover, AI-driven information retrieval systems prioritize contextually relevant content personalized to users' needs. This means that businesses must invest in creating content tailored to specific audiences and offering a seamless and engaging user experience.

4.2 Personalization and User Engagement

AI models can deliver highly personalized content based on user preferences, behavior, and history. This means that SEO strategies may need to focus more on creating content that engages users personally. Personalized content can help build stronger connections with users, increase engagement, and drive better results in an AI-driven search landscape.

The rise of AI-driven personalization could lead to a more fragmented content landscape, where businesses must create multiple versions of content tailored to different audiences and contexts. This could involve using AI to analyze user data and generate personalized content recommendations that align with individual preferences and behaviors.

Moreover, businesses must focus on creating engaging and interactive content, as AI models will likely prioritize content that keeps users engaged and encourages them to act. This could involve creating content that encourages user participation, such as quizzes, polls, and interactive videos, and leveraging AI to create personalized content experiences that adapt to the user's needs in real time.

4.3 The Integration of AI in Web Design

Web design must adapt as AI becomes more integrated into the user experience. This could involve creating more interactive, dynamic web pages that facilitate AI-driven interactions. For example, websites could feature AI-powered chatbots, providing users instant answers to their questions and guiding them through the content. Additionally, web pages may need to be structured to make extracting and presenting relevant information more accessible for AI.

Integrating AI in web design could create more personalized and adaptive websites that respond to user behavior in real time. This could involve using AI to analyze user data, generate personalized content recommendations, and create dynamic web pages that adapt to the user's preferences and behavior.

Moreover, businesses must focus on creating web designs optimized for AI-driven information retrieval, ensuring that content is structured to make it easy for AI to extract and present relevant information. This could involve using structured data and schema markup to provide context and metadata for AI models and optimizing content for different devices and platforms.

4.4 Enhancing Accessibility and Inclusivity

AI-driven information retrieval has the potential to enhance accessibility by providing content in multiple formats and languages. In the future, SEO strategies may need to prioritize accessibility, ensuring that content is available to a diverse range of users, including those with disabilities. This could involve creating content compatible with screen readers, providing alternative text for images, and offering multilingual options to cater to a global audience.

The rise of AI-driven information retrieval could also lead to the creation of more inclusive content experiences catering to diverse users. This could involve using AI to analyze user data and generate content recommendations that are tailored to different cultural contexts and preferences, as well as creating content that is accessible to users with disabilities.

Moreover, businesses will need to focus on creating content that is inclusive and representative of diverse perspectives, as AI models are likely to prioritize content that aligns with the values and preferences of diverse audiences. This could involve creating content that reflects the experiences and perspectives of different communities and ensuring that content is accessible to users with different needs and preferences.

CHAPTER

FIVE

The Future of Link Building and Backlinks

5.1 The Changing Role of Backlinks

Backlinks have long been a cornerstone of traditional SEO, critical in determining a website's authority and relevance. However, in a world where AI models like ChatGPT dominate information retrieval, the importance of backlinks may remain the same. AI systems are designed to prioritize content based on its relevance and quality rather than relying solely on the number of backlinks.

This shift could lead to a decline in the importance of traditional link-building strategies as businesses focus more on creating high-quality content optimized for AI-driven information retrieval. However, backlinks could still play a role in signaling the credibility and authority of content to AI models, mainly if they come from reputable and authoritative sources.

Moreover, the role of backlinks could evolve in an AI-driven SEO landscape, with a greater emphasis on the quality and relevance of backlinks rather than the quantity. This could involve building relationships with authoritative websites and influencers and creating content that naturally attracts high-quality backlinks from reputable sources.

5.2 Quality Over Quantity

If backlinks become less critical in an AI-driven SEO landscape, the focus may shift from quantity to quality. Instead of pursuing as many backlinks as possible, content creators and marketers may need to prioritize obtaining high-quality backlinks from authoritative and reputable sources. These high-quality backlinks could still play a role in signaling content credibility to AI models, even if they are not the primary factor in ranking.

Focusing on quality over quantity could lead to a more selective and strategic approach to link building, where businesses prioritize building relationships with authoritative websites and influencers in their industry. This could involve creating high-quality content that naturally attracts backlinks from reputable sources and engaging in digital PR and outreach efforts to build relationships with key influencers and media outlets.

Moreover, businesses must focus on creating valuable and relevant content for their target audiences. AI models will likely prioritize content that aligns with user intent and provides a positive user experience. This could involve creating in-depth and comprehensive content that addresses specific user needs and leveraging data and insights to create content that resonates with target audiences.

5.3 The Role of Social Proof and User Signals

Social proof and user signals could become more critical without traditional backlink-driven SEO. AI models like ChatGPT could consider factors such as user engagement, social shares, and reviews when determining the relevance and quality of content. This shift could lead to a greater emphasis on building a robust online presence, fostering user engagement, and encouraging positive user feedback.

The rise of social proof and user signals could lead to a more user-centric approach to SEO, where businesses focus on creating content that resonates with their audiences and encourages user participation. This could involve leveraging social media and online communities to build relationships with users and creating content that encourages social sharing and positive user feedback.

Moreover, businesses must focus on creating engaging and interactive content, as AI models will likely prioritize content that keeps users engaged and encourages them to act. This could involve creating content that encourages user participation, such as quizzes, polls, and interactive videos, and leveraging AI to create personalized content experiences that adapt to the user's needs in real time.

5.4 The Evolution of Digital PR and Outreach

As the role of backlinks evolves, digital PR and outreach strategies may also need to adapt. Rather than focusing solely on link-building campaigns, digital PR efforts could prioritize building relationships with influencers, securing media coverage, and creating content that resonates with target audiences. The goal would be to generate buzz and social proof, which could influence how AI models perceive and prioritize content.

The evolution of digital PR and outreach could lead to a more holistic approach to content marketing, where businesses focus on building relationships with key influencers and media outlets in their industry. This could involve creating high-quality content that naturally attracts media coverage and social sharing and engaging in digital PR efforts to build relationships with key influencers and thought leaders.

Moreover, businesses must focus on creating valuable and relevant content for their target audiences. AI models prioritize content that aligns with user intent and provides a positive user experience. This could involve creating in-depth and comprehensive content that addresses specific user needs and leveraging data and insights to create content that resonates with target audiences.

CHAPTER

SIX

Ethical Considerations and Challenges

6.1 The Ethics of AI-Generated Content

The rise of AI in content creation and information retrieval raises critical ethical questions. One key concern is the potential for AI-generated content to spread misinformation or biased information. Since AI models rely on the data they are trained on, they can inadvertently perpetuate existing biases or produce inaccurate information if the training data is flawed.

Content creators and marketers must navigate these ethical challenges carefully, ensuring that AI-generated content is accurate, unbiased, and aligned with ethical standards. This may involve implementing rigorous fact-checking processes, using diverse datasets, and being transparent about using AI in content creation.

Moreover, the ethical considerations surrounding AI-generated content could lead to developing new guidelines and best practices for content creators and marketers. This could involve creating ethical frameworks for using AI in content creation and establishing standards for accuracy, transparency, and accountability in AI-generated content.

6.2 The Impact on Privacy

AI models like ChatGPT can provide highly personalized content. Still, this level of personalization often requires access to vast amounts of user data, raising concerns about user privacy and data security. As AI-driven information retrieval becomes more prevalent, companies will need to balance the benefits of personalization with the need to protect user privacy.

Regulatory frameworks like the General Data Protection Regulation (GDPR) in Europe may significantly shape how AI models collect and use data. SEO strategies may need to account for these regulations, ensuring that content is practical and compliant with privacy laws.

Moreover, the rise of AI-driven personalization could lead to new challenges for businesses in managing user data and protecting privacy. This could involve implementing robust data security measures, being transparent with users about their data use, and ensuring that users have control over their data.

6.3 The Digital Divide

AI-driven information retrieval could exacerbate the digital divide, particularly if specific populations need more access to the necessary technology or digital literacy skills. Content creators and marketers must consider how to make their content accessible to a broad audience, regardless of technological barriers.

Efforts to bridge the digital divide involve creating content that is easy to understand and navigate, offering multiple formats to cater to different needs, and providing educational resources to help users engage with AI-driven technologies.

Moreover, businesses will need to focus on creating content that is inclusive and representative of diverse perspectives, as AI models are likely to prioritize content that aligns with the values and preferences of diverse audiences. This could involve creating content that reflects the experiences and perspectives of different communities and ensuring that content is accessible to users with different needs and preferences.

CHAPTER

SEVEN

Preparing For The Future Of SEO

7.1 Adapting to AI-Driven SEO

Businesses and content creators must adapt their strategies proactively to prepare for the future of SEO in a world dominated by AI. This may involve investing in AI tools and technologies, staying informed about the latest developments in AI-driven information retrieval, and experimenting with new content creation and optimization approaches.

The rise of AI-driven information retrieval could lead to a more competitive landscape where businesses must differentiate themselves by creating content that stands out in quality, originality, and relevance. This could involve leveraging AI to optimize content for different platforms and audiences and creating personalized content experiences that engage users on a deeper level.

Moreover, businesses must build relationships with their audiences, as AI models will likely prioritize content that resonates with users and encourages positive engagement. This could involve creating content that aligns with user intent, provides a positive user experience, and leverages social proof and user signals to build trust and credibility.

The latest developments in AI-driven information retrieval and experimenting with new content creation and optimization approaches.

The rise of AI-driven information retrieval could lead to a more competitive landscape where businesses must differentiate themselves by creating content that stands out in quality, originality, and relevance. This could involve leveraging AI to optimize content for different platforms and audiences and creating personalized content experiences that engage users on a deeper level.

Moreover, businesses must build relationships with their audiences, as AI models will likely prioritize content that resonates with users and encourages positive engagement. This could involve creating content that aligns with user intent, provides a positive user experience, and leverages social proof and user signals to build trust and credibility.

7.2 Embracing Innovation

The shift towards AI-driven SEO presents an opportunity for innovation. Companies willing to embrace change and explore new possibilities will be better positioned to succeed in this evolving landscape. This could involve experimenting with AI-generated content, exploring new content formats, and leveraging AI to enhance user experiences.

The rise of AI-driven information retrieval could lead to developing new content formats and experiences optimized for AI-driven interactions. This could involve creating interactive content that adapts to user preferences in real-time and leveraging emerging technologies like augmented reality (AR) and virtual reality (VR) to create immersive content experiences.

Moreover, businesses must focus on creating engaging and interactive content, as AI models will likely prioritize content that keeps users engaged and encourages them to act. This could involve creating content that encourages user participation, such as quizzes, polls, and interactive videos, and leveraging AI to create personalized content experiences that adapt to the user's needs in real-time.

7.3 Continuous Learning and Skill Development

As AI continues to evolve, so will the skills required to succeed in SEO. Content creators, marketers, and SEO professionals must continuously learn to stay ahead of the curve. This could involve learning about AI and machine learning, developing skills in data analysis, and staying up-to-date with the latest trends in digital marketing.

The rise of AI-driven information retrieval could lead to new opportunities for skill development and professional growth in the SEO industry. This could involve learning to use AI tools and technologies to optimize content for different platforms and audiences and developing expertise in data analysis, user experience design, and content strategy.

Moreover, businesses must focus on creating a continuous learning and innovation culture where employees are encouraged to experiment with new ideas and approaches to content creation and optimization. This could involve investing in training and development programs and fostering a collaborative, cross-disciplinary approach to SEO and digital marketing.

7.4 Collaboration and Cross-Disciplinary Approaches

The future of SEO will likely require collaboration between different disciplines, including AI, content creation, user experience design, and data analytics. Companies that foster a collaborative environment and encourage cross-disciplinary approaches will be better equipped to navigate the complexities of AI-driven SEO.

The rise of AI-driven information retrieval could lead to a more collaborative approach to content creation and optimization, where businesses work together to create personalized content experiences that engage users on a deeper level. This could involve leveraging AI to analyze user data, generate content recommendations that align with individual preferences and behaviors, and create content optimized for different platforms and audiences.

Moreover, businesses must build relationships with their audiences, as AI models will likely prioritize content that resonates with users and encourages positive engagement. This could involve creating content that aligns with user intent, provides a positive user experience, and leverages social proof and user signals to build trust and credibility.

CHAPTER

EIGHT

Conclusion

The potential rise of AI models like ChatGPT as the primary information retrieval method represents a significant shift in the digital landscape. If traditional search engines were to decline in favor of AI-driven systems, the future of SEO would undergo profound changes. While this shift presents challenges, it offers new opportunities for innovation, creativity, and user engagement.

To succeed in this evolving landscape, businesses, content creators, and SEO professionals must adapt their strategies, embrace AI technologies, and focus on creating high-quality, personalized content that resonates with users. By staying informed, investing in continuous learning, and fostering collaboration, the digital community can navigate the future of SEO in a world where AI plays a central role in shaping how information is retrieved and consumed.

In conclusion, while the future of SEO in an AI-dominated world may be uncertain, one thing is clear: the principles of relevance, quality, and user-centric content will remain at the core of successful digital strategies. By understanding and preparing for the potential changes brought about by AI, the digital community can continue to thrive in an ever-evolving landscape.

GLOSSARY OF TERMS

1. AI (Artificial Intelligence)

- **Definition**: A branch of computer science focused on creating systems capable of performing tasks that typically require human intelligence, such as understanding natural language, recognizing patterns, and making decisions.
- **Context in Booklet**: AI models like ChatGPT retrieve and generate information that mimics human conversation and decision-making processes.

2. ChatGPT

- **Definition**: A conversational AI model developed by OpenAI, designed to understand and generate human-like text based on the input it receives. It is an example of a large language model.
- **Context in Booklet**: ChatGPT is explored as a potential alternative to traditional search engines, capable of providing direct answers to user queries.

3. NLP (Natural Language Processing)

- **Definition**: A field of AI focusing on the interaction between computers and humans through natural language. It involves the ability of a machine to understand, interpret, and generate human language.

- **Context in Booklet**: NLP is a crucial technology behind AI models like ChatGPT, enabling them to process and generate contextually relevant text to user queries.

4. SEO (Search Engine Optimization)

- **Definition**: The practice of optimizing online content to improve its visibility and ranking on search engines, making it easier for users to find.
- **Context in Booklet**: SEO is discussed in the context of how it might evolve if AI models replace traditional search engines.

5. E-A-T (Expertise, Authoritativeness, Trustworthiness)

- **Definition**: A set of criteria used by search engines like Google to assess the quality and reliability of content. High E-A-T scores typically result in better search engine rankings.
- **Context in Booklet**: E-A-T is a potentially critical factor in AI-driven SEO, where AI models may prioritize content from authoritative and trustworthy sources.

6. Backlink

- **Definition**: A link from one website to another. Search engines often use backlinks to determine the authority and relevance of a website.
- **Context in Booklet**: The Booklet discusses the changing role of backlinks in SEO, as AI models may prioritize content quality over the number of backlinks.

7. PageRank

- **Definition**: An algorithm developed by Google to rank web pages in their search engine results. It measures the importance of web pages based on the number and quality of links pointing to them.
- **Context in Booklet**: PageRank is highlighted as one of the critical innovations that revolutionized search engine algorithms.

8. Keyword Density

- **Definition**: The percentage of times a keyword appears in a piece of content relative to the total number of words. It is used in SEO to optimize content for specific search terms.
- **Context in Booklet**: The Booklet discusses the potential decline in the importance of keyword density in favor of contextually relevant content aligned with user intent.

9. Long-Tail Keywords

- **Definition**: More specific and usually longer keyword phrases that are less commonly searched but often have a higher conversion value.
- **Context in Booklet**: Long-tail keywords may become more critical in an AI-driven SEO landscape, as AI models can interpret complex and natural language queries.

10. Structured Data

- **Definition**: Data that is organized in a specific format, often using schema markup, to help search engines understand the content of a web page.
- **Context in Booklet**: Structured data may become increasingly important for ensuring that AI models can

accurately interpret and retrieve relevant information from web pages.

11. Schema Markup

- **Definition**: A form of microdata that creates an enhanced description (often known as a rich snippet) that appears in search results.
- **Context in Booklet**: Schema markup is mentioned to provide AI models with context and metadata that can improve information retrieval.

12. Social Proof

- **Definition**: The concept that the actions and opinions of others influence people. In digital marketing, social proof can include customer reviews, testimonials, and social media shares.
- **Context in Booklet**: Social proof is discussed as a potential factor that AI models might consider when determining the relevance and quality of content.

13. User Signals

- **Definition**: Behavioral data from users that indicates how they interact with a website, such as click-through rates, time on page, and bounce rates.
- **Context in Booklet**: User signals could become important metrics for AI models in assessing the quality and relevance of content.

14. Conversational AI

- **Definition**: A type of AI designed to engage in human-like conversations, often used in chatbots and virtual assistants.
- **Context in Booklet**: Conversational AI is exemplified by models like ChatGPT, which may influence how users search for and receive information online.

15. Digital PR (Public Relations)

- **Definition**: Using online methods to enhance brand visibility and reputation, often through media coverage, influencer relationships, and content marketing.
- **Context in Booklet**: The evolution of digital PR concerns the changing importance of backlinks and the need to build relationships with influencers and media outlets.

16. Augmented Reality (AR)

- **Definition**: An interactive experience where digital information overlaps the physical world, often through smartphones or AR glasses.
- **Context in Booklet**: AR is mentioned as an emerging technology that could be leveraged to create new, immersive content experiences in an AI-driven SEO landscape.

17. Virtual Reality (VR)

- **Definition**: A fully immersive digital environment that users can interact with, typically through VR headsets.
- **Context in Booklet**: VR is discussed as a potential content format that could be optimized for AI-driven interactions and user engagement.

18. General Data Protection Regulation (GDPR)

- **Definition**: A regulation in EU law on data protection and privacy also addresses transferring personal data outside the EU.
- **Context in Booklet**: GDPR is mentioned in the context of AI-driven personalization and the need to balance personalization with user privacy.

19. Digital Divide

- **Definition**: The gap between individuals who have access to modern information and communication technology and those who do not, often due to socioeconomic, geographic, or educational factors.
- **Context in Booklet**: The digital divide is a potential challenge in ensuring that AI-driven information retrieval is accessible to all users.

20. Personalization

- **Definition**: Tailoring content and experiences to individual users based on their preferences, behavior, and history.
- **Context in Booklet**: Personalization is a crucial feature of AI-driven information retrieval, allowing for more relevant and engaging user experiences.

Made in United States
Cleveland, OH
05 November 2026

47642857R00026